Mission CRS-31

SpaceX's Next Leap in Space Exploration

A Compelling Breakdown of NASA's 31st Resupply Operation, Scientific Experiments, and Spacecraft Innovations

Joe E. Grayson

Copyright © 2024 Joe E. Grayson, All rights reserved.

No part of this publication may be reproduced, distributed, or transmitted in any form or by any means, including photocopying, recording, or other electronic or mechanical methods, without the prior written permission of the publisher, except in the case of brief quotations embodied in critical reviews and certain other noncommercial uses permitted by copyright law.

Table of Contents

Table of Contents

Introduction

Chapter 1: The Journey Begins – Preparations and Pre-Launch

Chapter 2: Liftoff and Ascent

Chapter 3: The Falcon 9 Rocket and Reusability

Chapter 4: The Dragon Spacecraft – A Modern Workhorse

Chapter 5: Critical Cargo and Scientific Payloads

Chapter 6: The Journey to the ISS – Orbital Maneuvers

Chapter 7: The ISS Reboost Capability Test

Chapter 8: The Role of Teamwork and Mission Control

Chapter 9: A Successful Return to Earth

Chapter 10: Reflections on Space Exploration and Future Implications

Conclusion

Introduction

SpaceX's 31st Cargo Resupply Mission, known as CRS-31, represents a significant milestone in the ongoing collaboration between SpaceX and NASA, aimed at maintaining and advancing the operations of the International Space Station (ISS). This mission builds upon a series of resupply efforts that have become essential to supporting both the scientific and logistical needs of the ISS, ensuring that the orbiting laboratory remains an active center for groundbreaking research and technological progress. By transporting essential supplies, scientific experiments, and equipment to the ISS, CRS-31 demonstrates the critical role private companies like SpaceX now play in furthering humanity's exploration and understanding of space.

The history of SpaceX's resupply missions began in 2012, marking a new chapter in space exploration. Through the NASA Commercial Resupply Services (CRS) program, SpaceX became the first private company to deliver cargo to the ISS. This achievement opened doors to a partnership that emphasizes efficiency, cost-effectiveness, and innovation. A defining feature of SpaceX's approach has been the development of reusable rockets, which drastically reduce the cost of space missions. The Falcon 9 rocket, used in these missions, is designed to return to Earth and be launched again, allowing for multiple missions with the same hardware. This reusability has not only made space exploration more accessible but has also set a new standard in the aerospace industry, inspiring other companies and agencies to pursue similar advancements.

The CRS-31 mission is designed with several key objectives, each contributing to the broader goals of scientific research, operational support, and technological innovation. First, this mission brings over three tons of cargo to the ISS, including scientific equipment and samples that will support studies on solar wind, plant growth in space, and material resilience in a microgravity environment. These experiments are crucial as they provide insights that could lead to advancements in deep-space exploration and innovations with applications on Earth. Additionally, CRS-31 includes essential supplies for the ISS crew, enhancing their comfort and ability to carry out their daily tasks.

Another groundbreaking aspect of CRS-31 is its experimental reboost capability. Typically, the ISS relies on specific spacecraft to maintain its altitude, compensating for the gradual loss caused by atmospheric drag. For the first time, Dragon will test its ability to perform this reboost maneuver, an essential function that could allow SpaceX to take on a more comprehensive role in station maintenance. If successful, this capability would demonstrate Dragon's potential for even greater utility in supporting the ISS and future orbital platforms.

As SpaceX and NASA continue their collaboration, CRS-31 stands as a testament to the progress made in commercial spaceflight and highlights the critical contributions of private companies in expanding humanity's reach beyond Earth. The mission not only provides immediate benefits to the ISS but also paves the way for future developments that could redefine our approach to living and working in space.

Chapter 1: The Journey Begins – Preparations and Pre-Launch

The preparation for the CRS-31 mission began long before the final countdown, with meticulous planning and coordination to ensure the mission's success. One of the first stages involved cargo preparation, where scientists and engineers carefully selected and organized the items that would travel aboard Dragon to the International Space Station. This cargo included over three tons of essential supplies, scientific equipment, and experimental materials. Each item was rigorously inspected and cataloged to guarantee that it would not only fit securely within Dragon's compartments but would also be accessible to the ISS crew once unloaded. This process required precise planning to maximize space and ensure that critical items, such as scientific samples and life-support materials, would be delivered intact and ready for use.

Simultaneously, teams at SpaceX were performing exhaustive checks on the Dragon spacecraft and the Falcon 9 rocket, verifying that every component was in optimal condition. These inspections covered everything from fuel lines and thrusters to guidance systems and thermal protection panels. Each part of the spacecraft underwent rigorous stress tests and simulations, mirroring the conditions Dragon would encounter during its journey to orbit. Engineers closely examined every system and subsystem, confirming that each could withstand the intense forces and temperatures that would come during liftoff, ascent, and re-entry. Final assembly was the culmination of these efforts, bringing together all components in preparation for launch. The Falcon 9 rocket was mated to the

Dragon spacecraft, and the fully integrated vehicle was positioned for final inspections, ready to stand as a single unit for its upcoming mission.

The chosen launch site for CRS-31 was Launch Complex 39A at NASA's Kennedy Space Center in Florida, a location steeped in the history of American space exploration. Launch Complex 39A has witnessed some of the most pivotal moments in space history, beginning with the Apollo program. It was from this very pad that Apollo 11 launched in 1969, sending astronauts to the Moon for the first time. Decades later, Complex 39A became the site for numerous space shuttle launches, including missions to the ISS and repairs of the Hubble Space Telescope. This iconic pad, symbolizing the spirit of human exploration and innovation, was adapted for modern missions through a lease between NASA and SpaceX. Today, it supports commercial space missions, symbolizing a new era where private and public partnerships propel humanity's journey into space.

For SpaceX, launching from Complex 39A underscores the continuation of groundbreaking achievements in space exploration. The pad's rich history now merges with the future-focused goals of commercial spaceflight, linking past accomplishments to the ambitions of today. In the final hours leading up to the CRS-31 launch, the vehicle stood poised at Complex 39A, embodying the legacy of its predecessors while marking yet another chapter in this storied launch site's evolution.

As the clock ticked down toward liftoff, the final countdown sequence for the CRS-31 mission brought a wave of activity and coordination between SpaceX and NASA teams. Hours before launch, engineers and mission controllers carried out a series of critical checks and verifications to ensure the Falcon 9 rocket and Dragon spacecraft were fully prepared for their journey to the International Space Station. The meticulous process involved verifying systems across the rocket and spacecraft, including propulsion, guidance, communication, and onboard computers. Every component was tested to confirm its readiness, with teams running through a rigorous checklist that left no detail overlooked.

Weather conditions were a crucial factor in the countdown. SpaceX's team closely monitored local atmospheric data to ensure that wind speeds, cloud cover, and lightning risks met the requirements for a safe launch. Even minor fluctuations in these parameters could impact the rocket's trajectory or the success of the launch, so meteorologists provided real-time updates, tracking any developing conditions that might delay or reschedule liftoff. Fortunately, as the countdown progressed, the weather remained favorable, giving mission control the green light to proceed.

Meanwhile, communication between SpaceX's control center in Hawthorne, California, and NASA's mission control at the Johnson Space Center in Houston, Texas, was constant. Both teams shared updates and synchronized their actions, ensuring that everyone was aligned on the status of the spacecraft and the progress of the countdown. This collaboration exemplified the teamwork required to manage a complex space mission, as each decision and check was carefully coordinated across multiple locations and teams. Engineers on the ground exchanged data, monitoring the Falcon 9's fuel levels, Dragon's telemetry systems, and the overall health of the vehicle.

At various stages in the countdown, mission controllers announced "go" or "no-go" statuses, evaluating each system's readiness in real time. These go/no-go calls allowed the teams to pause if any irregularities arose, addressing issues immediately or making adjustments if needed. One by one, each system was cleared, and the final moments of the countdown approached. With minutes remaining, Falcon 9's first and second stages were fully loaded with nearly one million pounds of fuel and liquid oxygen. The rocket's autonomous flight computer took control, assuming command of the final launch procedures.

In the last moments, all eyes were on the Falcon 9 as it stood ready on the pad. The teams gave a final "go" for launch, confirming that all systems were nominal and the mission was ready to proceed. As the countdown hit zero, engines ignited in a fiery burst, and the rocket ascended, marking the beginning of CRS-31's journey to the ISS. This sequence of checks, verifications, and precise timing underscored the dedication, expertise, and focus required to bring each mission to fruition, as every step in the countdown brought SpaceX and NASA one step closer to fulfilling their shared vision of space exploration.

Chapter 2: Liftoff and Ascent

The launch of SpaceX's CRS-31 mission marked a meticulously choreographed sequence of events that demonstrated the precision and power behind modern space exploration. As the countdown reached zero, the Falcon 9 rocket's engines ignited in a controlled burst, lifting the rocket off the pad at Kennedy Space Center's Launch Complex 39A. In those initial moments, Falcon 9 soared upward, leaving a trail of flame and exhaust as it cleared the tower and accelerated into the skies. Within seconds, the rocket's speed increased, quickly pushing it past the sound barrier and into supersonic flight. This transition to supersonic speeds occurred smoothly, and Falcon 9's ascent continued with the guidance of ground-based telemetry, which relayed real-time data back to SpaceX's mission control.

Falcon 9's ascent was designed to maximize efficiency, with the rocket following a precise trajectory calculated to conserve fuel while optimizing speed. As the rocket climbed higher, the atmospheric drag decreased, allowing the engines to maintain high thrust with minimal resistance. Falcon 9's first stage, which is equipped with nine Merlin engines, provided the initial thrust needed to push the vehicle out of Earth's lower atmosphere. Once it reached a predetermined altitude, the rocket prepared for one of the most crucial steps in the mission: stage separation.

Stage separation is a pivotal maneuver in multi-stage rockets like Falcon 9, as it allows each section to fulfill its role efficiently. After completing its main engine burn, the first stage shut down, effectively ending its part of the journey. At this point, the first and second stages detached, allowing the now-spent first stage to begin its descent back to Earth, where it would attempt a controlled landing for potential reuse. Meanwhile, the second stage ignited its single Merlin Vacuum engine, designed specifically to operate in the vacuum of space. This second stage burn propelled the Dragon spacecraft even higher, pushing it into low Earth orbit.

With the first stage returned safely to Earth, all focus turned to the second stage, which continued to carry Dragon toward its target orbit. The second stage's role was to provide the final push, ensuring that Dragon reached the precise altitude and trajectory needed for its autonomous journey to the International Space Station. Once Dragon was confirmed in its intended orbit, the second stage separated, leaving Dragon to complete the remainder of its journey using its own onboard navigation and propulsion systems. This sequence of launches and separations underscored the sophistication of Falcon 9's design, showcasing how each stage's specialized function contributed to the success of the mission and ensured Dragon's safe delivery to space.

Once Falcon 9 and Dragon were on their way to orbit, real-time flight monitoring became the essential lifeline between the spacecraft and mission control. Telemetry updates flowed constantly, transmitting a stream of data from the rocket's sensors back to SpaceX's ground teams. These updates included information on altitude, speed, fuel levels, engine temperatures, pressure readings, and trajectory alignment. Every fraction of a second, Falcon 9's systems provided thousands of data points, allowing engineers and controllers to track the vehicle's exact status and performance.

During the ascent, telemetry ensured that each stage of Falcon 9's journey followed the planned trajectory. The data enabled controllers to confirm the smooth operation of key events, such as the transition to supersonic speeds, the main engine cutoff, and the precise timing of stage separation. Any slight deviation from expected parameters would trigger alerts, allowing the ground teams to assess the situation and, if necessary, adjust or abort the mission to protect the spacecraft and crew on the ISS.

Telemetry wasn't limited to tracking Falcon 9 alone. Once Dragon separated and entered its own orbit, its systems began sending similar updates to mission control. Dragon's onboard navigation, powered by an array of sensors and communication systems, relayed its orientation, propulsion status, and approach path toward the ISS. With real-time feedback on Dragon's exact position and alignment, the teams could monitor each maneuver and ensure that Dragon stayed on course. These tracking efforts continued as Dragon performed a series of planned burns, gradually raising its orbit and preparing for its eventual rendezvous with the ISS.

In addition to SpaceX's monitoring, NASA's mission control in Houston was also involved in tracking Dragon's progress. The collaboration allowed both SpaceX and NASA to cross-verify data, strengthening the mission's safety and accuracy. Each piece of information shared between the control centers provided mission controllers with a comprehensive view of the spacecraft's journey, helping them anticipate and respond to any challenges that might arise.

This constant stream of data formed the backbone of mission assurance, with real-time telemetry making it possible to closely oversee every aspect of Falcon 9 and Dragon's journey. The insights gained from telemetry, tracking, and continuous monitoring allowed the teams to confidently confirm that Dragon was safely on its way to deliver essential supplies, experiments, and equipment to the ISS.

Chapter 3: The Falcon 9 Rocket and Reusability

The Falcon 9 rocket stands as a technical marvel in modern spaceflight, designed with a focus on both power and efficiency to deliver payloads into orbit. Its two-stage structure is essential to its performance, enabling the rocket to conserve fuel and optimize each stage for a specific function. The first stage houses nine Merlin engines arranged in a unique configuration known as the Octaweb. This layout provides stability and redundancy, allowing Falcon 9 to generate the thrust needed to lift heavy payloads through Earth's atmosphere. At full power, these engines produce 1.7 million pounds of thrust, which pushes the rocket to supersonic speeds within seconds of liftoff.

The second stage, equipped with a single Merlin Vacuum engine, takes over after the first stage separates and completes the critical task of placing the payload in its designated orbit. This engine is optimized for performance in the vacuum of space, operating with high efficiency to propel the Dragon spacecraft to the International Space Station. The two-stage design allows Falcon 9 to maximize fuel efficiency, as the first stage carries the initial burden of lift-off and ascent, while the lighter second stage completes the journey without the weight of the lower stage.

A defining characteristic of Falcon 9 is its commitment to reusability, a feature that has reshaped the economics of space travel. SpaceX has designed the first stage of Falcon 9 to return safely to Earth after launch, where it can be recovered, inspected, and reused for future missions. This ability to reuse hardware represents a major leap forward in making space missions more sustainable and cost-effective. Traditionally, rockets have been single-use vehicles, with each launch requiring an entirely new build. However, Falcon 9's reusability enables SpaceX to launch more frequently at a fraction of the cost, as refurbishing a used stage is significantly less expensive than constructing a new one.

The sustainability that reusability brings has allowed SpaceX to conduct multiple missions with the same booster, some flown up to ten times. This approach not only reduces costs for SpaceX but also brings down prices for customers and partners, making access to space more affordable and accelerating the pace of space exploration. The reusability model pioneered by Falcon 9 has inspired other companies and space agencies to consider similar designs, potentially leading to a new standard in rocketry that balances innovation with economic efficiency.

Through its two-stage design and emphasis on reusability, Falcon 9 has transformed expectations for what is possible in spaceflight. It combines power and precision with sustainability, making space more accessible and laying the groundwork for a future where routine space missions can be achieved with minimal waste and maximum efficiency.

The recovery of Falcon 9's first stage is one of the most innovative aspects of SpaceX's approach to space travel, representing a milestone in reusability and cost-efficiency. After completing its task of propelling the rocket through the atmosphere and reaching the desired altitude, the first stage shuts down its engines and separates from the second stage, which continues to carry the payload into orbit. At this point, the first stage begins its journey back to Earth, guided by a carefully orchestrated sequence of maneuvers designed to ensure a safe and precise landing.

The first of these maneuvers is the boost-back burn, a critical step in redirecting the stage toward its designated landing site. This burn is achieved by relighting three of the nine Merlin engines, which briefly fire to alter the stage's trajectory. By adjusting its path, the first stage can either return to a landing zone near the launch site or aim for an autonomous drone ship stationed in the ocean. This choice depends on the mission's specific requirements and the amount of fuel remaining after launch.

Following the boost-back burn, the first stage performs a controlled descent through the atmosphere. To slow its re-entry and protect it from the extreme heat generated during descent, the stage initiates an entry burn. This burn involves igniting three engines once again to reduce the vehicle's speed, minimizing stress on the stage's structural integrity and allowing for a safer approach to the landing area. The Falcon 9's grid fins, which extend from the sides of the stage, also play a crucial role during this descent, helping to stabilize and steer the rocket as it approaches its landing target.

Finally, as the first stage nears its landing site, it executes the landing burn. This final burn relies on a single engine, which provides the precise amount of thrust required to bring the rocket to a soft landing. In the moments before touchdown, the landing legs deploy, allowing the stage to safely settle on solid ground or the deck of a drone ship, completing the recovery process. Successfully recovering the first stage allows SpaceX to inspect and refurbish it for future flights, a breakthrough in reducing the cost of access to space.

The ability to reuse the first stage has far-reaching implications for the future of space exploration. Traditionally, rockets have been single-use, with each launch requiring a new vehicle, which adds significantly to the expense and environmental impact of space missions. By reusing the Falcon 9's first stage, SpaceX has drastically lowered the cost of launches, enabling more frequent missions and expanding opportunities for scientific research, commercial ventures, and exploratory missions. This reusability model not only makes space travel more accessible but also serves as a prototype for future rockets that could potentially bring humanity closer to sustainable space travel.

With every successful recovery, SpaceX continues to refine its techniques, moving closer to the goal of rapid and reliable reuse. This approach has inspired a shift in the aerospace industry, encouraging other companies and agencies to pursue reusable designs, which could further reduce the cost and increase the accessibility of space exploration. The Falcon 9 first stage recovery stands as a landmark achievement, underscoring SpaceX's commitment to a future where routine, sustainable space missions are a reality.

Chapter 4: The Dragon Spacecraft – A Modern Workhorse

The Dragon spacecraft, a pioneering creation by SpaceX, is designed to transport cargo safely to and from the International Space Station (ISS) with remarkable efficiency and reliability. Built with both pressurized and unpressurized sections, Dragon's structure is optimized to carry a wide variety of supplies, scientific equipment, and experimental materials essential to the ISS mission.

The pressurized section of Dragon serves as a protected area for sensitive cargo, such as scientific samples, food supplies, and any items that need a controlled environment. This section is fully enclosed and climate-controlled, ensuring that temperature-sensitive materials and delicate equipment remain stable and secure during launch, travel, and docking. Crew members on the ISS can easily access this section to retrieve cargo once Dragon is docked, allowing for swift and organized unloading.

Beneath the pressurized capsule lies Dragon's unpressurized cargo trunk, an open section designed to carry bulkier items that do not require environmental control. This trunk allows Dragon to transport larger, non-sensitive equipment, such as replacement parts or external scientific instruments intended for installation on the ISS's exterior. The versatility of Dragon's cargo sections enables it to accommodate a wide range of mission needs, making it an invaluable asset for the ISS and other potential space missions in the future.

Dragon's autonomous navigation capabilities allow it to maneuver independently as it approaches the ISS. This complex task is made possible by a combination of advanced guidance systems and the use of Draco thrusters. These Draco thrusters, mounted around the spacecraft, are small but powerful engines that allow Dragon to perform precise adjustments in orientation and speed. With a thrust of around 90 pounds per engine, the Draco thrusters enable the spacecraft to carry out delicate maneuvers, such as alignment corrections and minor course adjustments.

As Dragon nears the ISS, its onboard navigation systems, which include a blend of GPS, radar, and optical sensors, work together to calculate its position relative to the station. These systems continuously relay data back to mission control, where engineers monitor Dragon's progress and can intervene if needed. However, Dragon is fully equipped to perform the docking sequence on its own, guided by pre-programmed instructions and real-time sensor feedback.

Once Dragon reaches a designated approach point near the ISS, it begins a series of pre-docking maneuvers, making slight adjustments to ensure it aligns perfectly with the station's docking port. These last steps require incredible precision, as the spacecraft must approach at a controlled, slow speed to avoid any sudden impact with the station. Upon reaching the docking port, Dragon's autonomous system completes the final attachment to the ISS, establishing a secure link that allows for the safe transfer of cargo.

The spacecraft's ability to autonomously navigate and dock without constant human control not only enhances mission efficiency but also frees up valuable time for astronauts and ground crew.

Dragon's navigation and docking features make it a reliable and resilient partner in resupply missions, allowing it to safely deliver essential materials to the ISS and support continuous research and exploration.

Dragon's cargo capacity is one of its standout features, engineered to support a variety of mission-critical needs for the International Space Station (ISS). With a payload capacity of up to 6,000 kilograms (around 13,228 pounds) in its most recent configurations, Dragon is capable of transporting a diverse range of cargo to sustain ISS operations and facilitate scientific research. This capacity is distributed between its pressurized and unpressurized sections, allowing it to carry both sensitive, protected items and larger equipment that can withstand the open environment of space.

Within the pressurized section, Dragon can safely transport scientific equipment and experimental materials that require stable, controlled conditions. For missions like CRS-31, these payloads often include biological samples, experimental modules, and other instruments that will be used by the ISS crew to conduct research in microgravity. These scientific packages may cover a broad spectrum of research, from examining plant growth in space to testing the resilience of materials exposed to space conditions. Each of these experiments is meticulously packed to ensure it arrives in optimal condition for immediate deployment upon reaching the station.

Dragon's pressurized cargo section also carries essential crew supplies, which are critical for the well-being and productivity of astronauts on board the ISS. These items include food, water, personal care products, and medical supplies that sustain crew members during their extended stays in space. Fresh food kits, sometimes containing fruits and vegetables, are also included in resupply missions, providing astronauts with nutritious options and a morale boost amid their daily routines. Additionally, Dragon can transport clothing and other personal items requested by the crew, allowing them to maintain a level of comfort and continuity in their isolated environment.

In the unpressurized cargo trunk, Dragon can carry large equipment, replacement parts, and even external scientific instruments intended for installation outside the ISS. This section is particularly valuable for transporting items like solar panels, research modules, and tools that can be installed directly onto the station's exterior. By accommodating unpressurized cargo, Dragon expands its mission flexibility, supporting not only the internal needs of the ISS but also its external infrastructure and capacity for scientific experiments conducted outside the protective environment of the station's modules.

The versatility of Dragon's cargo configuration makes it a linchpin in ISS resupply missions, as it can accommodate the station's dynamic and varied needs. This ability to carry a high volume of cargo, spanning scientific, logistical, and personal items, ensures that the ISS remains well-stocked, fully operational, and primed for new research. As the only U.S. spacecraft capable of round-trip journeys with significant cargo capacity, Dragon stands as an essential partner in sustaining the ISS and advancing our understanding of life and science in space.

Chapter 5: Critical Cargo and Scientific Payloads

The CRS-31 mission carried a diverse array of payloads to the International Space Station, designed to support the daily needs of the crew while advancing scientific research. Among the items packed into Dragon's cargo sections were essential supplies for the astronauts, including food, water, and other personal necessities required for their long-duration mission. These crew essentials covered everything from basic sustenance to comfort items, ensuring that the astronauts could maintain their well-being and focus on their work aboard the station. In addition to these basics, the CRS-31 mission included a special holiday kit, which brought a bit of festivity to the crew's lives with treats like fresh fruits, vegetables, and even gourmet items such as lobster and quail. These holiday additions provided a morale boost, giving the crew a chance to enjoy something familiar and celebratory despite their isolation in space.

Beyond these necessities and comforts, the CRS-31 mission also brought a range of scientific payloads, each aimed at expanding our knowledge across multiple fields of study. One of the mission's primary scientific investigations involved a study of solar wind, the stream of charged particles constantly released from the Sun's outer layers. This study seeks to gather data to validate theories about the solar wind's behavior, which remains an area of active research. Understanding solar wind dynamics is critical for future space exploration, as these particles can pose significant risks to spacecraft and astronauts. The data collected will contribute to creating safer travel strategies for missions venturing beyond Earth's magnetic protection.

Another intriguing experiment involved the growth and resilience of arctic moss in space. Arctic moss is a plant uniquely suited to extreme conditions on Earth, and scientists hope to gain insights into how it might behave and adapt in the microgravity environment of space. By observing the moss's response to space conditions, researchers aim to identify biological systems that might be suitable for life-support applications in long-term space missions. Success in this area could inform the development of sustainable plant-based systems for producing oxygen, recycling waste, and even providing fresh food on future missions to destinations like Mars.

The CRS-31 mission also included materials science experiments aimed at understanding the effects of prolonged exposure to the harsh space environment. By subjecting various materials to the vacuum, radiation, and temperature extremes outside the ISS, scientists can evaluate their durability and resistance to degradation. This research is critical for designing more robust spacecraft, satellites, and space equipment that can withstand the unique challenges of space over extended periods. The materials exposure tests may yield findings that improve the safety and longevity of future space missions, as well as provide insights into developing materials with applications on Earth, especially in industries where durability and resilience are paramount.

These diverse payloads underscore the multifaceted goals of the CRS-31 mission. By supporting both the immediate needs of the crew and the long-term objectives of space exploration, the mission's cargo highlights the essential role that resupply missions play in advancing human presence in space. Every item, from holiday treats to scientific instruments, contributes to a broader understanding of how humanity can live, work, and eventually thrive beyond our home planet.

The experiments aboard the CRS-31 mission have far-reaching implications, both for humanity's aspirations in deep-space exploration and for technological advancements that can benefit life on Earth. Each scientific investigation conducted on the International Space Station (ISS) provides insights that push the boundaries of what is possible in space, while simultaneously fostering innovations that can be applied to everyday challenges on our home planet.

The solar wind study, for example, addresses a phenomenon that affects every spacecraft and astronaut leaving the protective shield of Earth's magnetic field. Solar wind, composed of high-energy particles from the Sun, can interfere with electronic equipment, threaten astronaut health, and impact the structural integrity of spacecraft. By studying the solar wind in detail, scientists aim to develop more accurate predictive models, which could lead to better shielding technologies and informed safety protocols for deep-space missions. This research is essential for future exploration to Mars and beyond, where understanding and mitigating solar radiation hazards will be crucial for mission success. Additionally, these insights may lead to improvements in satellite technology, ensuring that vital communication and weather-monitoring satellites around Earth are better protected from solar events.

The arctic moss experiment represents an important step in understanding how biological systems might sustain human life in space. Arctic moss, known for its resilience in extreme Earth environments, is being studied to determine its adaptability to the conditions aboard the ISS. If arctic moss thrives in microgravity, it could serve as a model for developing plant-based life support systems that provide oxygen, recycle carbon dioxide, and even produce fresh food on long-term space missions. Such systems would be invaluable for crewed missions to Mars or lunar habitats, where the capacity for self-sustenance could reduce dependency on Earth-based resupply missions. Moreover, findings from this study could inspire advances in sustainable agricultural practices on Earth, particularly in areas with harsh climates where traditional crops struggle to survive.

The materials exposure tests conducted on this mission are equally critical. Space is a hostile environment where materials are exposed to extreme radiation, temperature fluctuations, and micro-meteoroid impacts. By studying how different materials respond to prolonged exposure, researchers can identify or develop materials that are more resilient, lightweight, and durable. These advancements could lead to the creation of longer-lasting spacecraft, enabling exploratory missions to distant planets without the constant need for repairs or replacements. On Earth, this knowledge has direct applications in fields like construction, automotive, and aerospace, where materials that withstand extreme conditions are highly valuable. Enhanced materials could also benefit industries like energy production, where infrastructure durability under extreme conditions is crucial.

Through these experiments, CRS-31 showcases the dual-purpose nature of ISS-based research: enabling humanity to explore and settle other worlds, while also driving practical innovations on Earth. As we push the limits of what we can achieve in space, the knowledge gained circles back to improve technology, safety, and sustainability in our daily lives. The discoveries made on this mission lay groundwork for a future where both Earth and space-based challenges can be addressed with shared solutions, paving the way for progress that bridges the gap between our planet and the vast reaches of space.

Chapter 6: The Journey to the ISS – Orbital Maneuvers

After reaching low Earth orbit, Dragon's journey to the International Space Station (ISS) requires a series of carefully planned orbital maneuvers called phasing burns. These five key phasing burns are performed using Dragon's Draco thrusters, each burn incrementally adjusting Dragon's orbit to align with the ISS's path. The goal is to gradually raise Dragon's orbit and synchronize its approach speed and trajectory so that it can safely rendezvous with the ISS. Each burn is calculated based on real-time data, which includes the ISS's position, Dragon's current orbit, and the necessary adjustments to bring them into alignment.

The first burn, often called the "Phase Burn," initiates the gradual orbit-raising process by increasing Dragon's altitude at the highest point of its initial orbit, called the apogee. This adjustment moves Dragon closer to the ISS's orbital altitude, bringing it one step nearer to the station's path. The second burn, the "Boost Burn," builds on this initial maneuver by further increasing Dragon's altitude at a calculated point in its orbit, refining its trajectory and creating a closer approach with the ISS's position.

As Dragon continues its orbit, a third maneuver, known as the "Transfer Burn," brings it into an orbit just below the ISS, positioning it roughly 10 kilometers lower than the station. This close proximity allows Dragon to approach the station without overtaking it, setting the stage for the final approach phase. Following this, a fourth burn, the "Coelliptic Burn," fine-tunes Dragon's orbit so it essentially mirrors the ISS's path, matching the spacecraft's speed and alignment at a stable distance just below the ISS.

The final burn, often called the "Final Coelliptic Burn," raises Dragon's orbit just enough to maintain a position approximately 2.5 kilometers below the ISS, aligning it for the last steps of the docking sequence. This set of five burns, executed with precise timing, sets Dragon on a safe trajectory for rendezvous, balancing proximity with controlled separation to prevent any unintentional drift or collision with the ISS.

As Dragon begins its approach, precision becomes paramount. The spacecraft's navigation system uses a combination of GPS, radar, and optical sensors to track its relative position and velocity as it nears the ISS. Small adjustments are continually made, supported by Dragon's Draco thrusters, which enable fine-tuned control for each movement. These onboard systems, guided by pre-programmed instructions, allow Dragon to inch closer to the ISS at a controlled, slow pace, ensuring the approach is safe and that the docking port remains in alignment.

In the final stages, Dragon must match its position with a docking port on the ISS down to millimeter-level accuracy. Each adjustment in orientation and speed is critical, as any misalignment could disrupt the docking sequence. Dragon's systems perform these maneuvers autonomously, but both NASA and SpaceX teams on Earth monitor the process closely, prepared to intervene if needed. The docking system, which includes sensors and locking mechanisms, ensures a secure connection once Dragon reaches the docking port.

This precise approach and docking capability highlights the sophistication of Dragon's design, demonstrating its ability to execute complex maneuvers that enable it to safely deliver cargo, supplies, and experiments to the ISS. The spacecraft's navigation and docking systems enable it to function autonomously, reducing demands on crew time and enhancing mission efficiency, all while ensuring a smooth and safe transfer of resources to the station.

Dragon's docking with the International Space Station (ISS) is a carefully orchestrated process that requires precision, coordination, and constant communication between the spacecraft and mission control teams at both NASA and SpaceX. As Dragon approaches the station after its series of phasing burns, it enters a "keep-out sphere," an imaginary boundary around the ISS that prevents any unscheduled or rapid movements within close range of the station. From this point onward, every maneuver is executed with extreme care to ensure the safety of both Dragon and the ISS.

Once Dragon is positioned within a safe distance from the ISS, it reaches a pre-determined waypoint known as "Waypoint 0," located about 400 meters below the station. Here, Dragon temporarily pauses, allowing mission control to conduct a thorough assessment of its alignment, position, and systems. SpaceX's mission control team in Hawthorne, California, and NASA's mission control at the Johnson Space Center in Houston, Texas, work in tandem, reviewing telemetry data, radar, and visual confirmation to verify that Dragon is precisely where it needs to be and functioning as expected.

With clearance granted, Dragon proceeds to "Waypoint 1," a closer position about 220 meters from the ISS. Here, another pause allows for additional checks to confirm that the spacecraft's path and orientation remain ideal for docking. This phased approach, with pauses at specific intervals, allows mission controllers to monitor Dragon's status and ensures that any potential issue can be addressed before it reaches the docking interface.

After all systems are confirmed as "go" at Waypoint 1, Dragon moves forward to "Waypoint 2," just 20 meters from the ISS docking port. Here, the spacecraft enters its final approach stage, slowing its speed to mere centimeters per second. This gradual approach is essential for preventing any undue stress on the station's structure or docking mechanisms. Using its optical sensors and navigation systems, Dragon autonomously adjusts its position, angle, and orientation, precisely aligning with the docking port.

Once in place, the docking system's sensors ensure Dragon's docking mechanism and the ISS's port are perfectly aligned. At this point, the spacecraft's docking rings initiate a "soft capture," where the ports connect and form an initial, lightly attached interface. This soft capture enables the systems to absorb any slight movements between the two structures, allowing them to stabilize before moving into the hard-docking phase. After soft capture, the docking system's latches engage, securing the connection firmly. This "hard capture" involves the activation of hooks and locks, which pull Dragon securely against the ISS docking port, creating a sealed, pressurized connection that allows crew members to access the cargo inside the spacecraft.

Throughout this process, SpaceX and NASA mission control teams maintain continuous communication, monitoring every step from Dragon's initial approach to the final lock-in. They ensure that all safety protocols are observed, and in case of any unexpected behavior, mission controllers have the ability to send override commands to pause or adjust the approach as needed. This shared oversight allows both organizations to respond swiftly to any changes, ensuring the docking process is completed without incident.

Once hard capture is confirmed, NASA astronauts aboard the ISS prepare to pressurize the vestibule—the space between Dragon and the ISS. They conduct leak checks to confirm that the connection is airtight, a crucial step for both the safety of the crew and the preservation of cargo integrity. After the area is fully pressurized and verified safe, the ISS crew members open the hatch, marking the official completion of Dragon's docking process. At this point, they can begin unloading the cargo, transferring the supplies, scientific experiments, and essential equipment delivered by Dragon into the ISS.

Dragon's docking is a seamless blend of advanced technology and skilled teamwork, embodying the precision and expertise of both SpaceX and NASA. This collaborative approach enables a safe and reliable transfer of vital resources, ensuring that the ISS remains fully operational and that new research can continue to advance humanity's understanding of life in space.

Chapter 7: The ISS Reboost Capability Test

Reboosting is a critical function that involves adjusting the International Space Station's (ISS) altitude to counteract the natural phenomenon of orbital decay. As the ISS orbits Earth, it gradually loses altitude due to the small but persistent drag caused by Earth's outer atmosphere. Even at the station's altitude of approximately 400 kilometers, atmospheric particles exert a subtle force on the ISS, causing it to slowly descend over time. Left unchecked, this gradual descent could bring the station dangerously close to the denser layers of the atmosphere, increasing the risk of re-entry and structural damage. Regular reboosts lift the station back to its proper orbit, helping to maintain a safe altitude and ensuring that it continues to function as an effective research platform.

Traditionally, reboosting has been performed by spacecraft such as Russia's Progress cargo vehicle or by using thrusters on the ISS's service modules. These vehicles periodically dock with the ISS and use their thrusters to push the station upward, counteracting the effects of orbital decay and maintaining its altitude. Reboosts are planned at intervals throughout the year to keep the station in its designated orbit, preserving the station's position for reliable operations and consistent research conditions.

Dragon's reboost experiment on the CRS-31 mission represents a new capability, testing whether SpaceX's cargo vehicle can take on this essential function. If successful, this reboost test would demonstrate that Dragon can independently adjust the ISS's altitude, adding flexibility to station maintenance and reducing reliance on other spacecraft. As part of this experiment, Dragon's Draco thrusters are programmed to perform a controlled burn, applying just the right amount of thrust to lift the ISS by a small but significant margin.

The significance of Dragon's reboost test extends beyond just maintaining altitude. Demonstrating reboost capabilities in Dragon would increase operational resilience for the ISS, as it would introduce a U.S.-based option for altitude adjustments. With Dragon potentially joining the list of vehicles capable of reboosting, mission planners could have greater flexibility in scheduling and executing these maneuvers. In the long term, having multiple vehicles equipped to perform reboosts contributes to the ISS's operational lifespan, making it easier to plan for continuous support even as certain partner vehicles, like Russia's Progress, undergo upgrades or changes in availability.

Additionally, data from Dragon's reboost test will inform future reboost capabilities and lay groundwork for future space missions. Spacecraft designed for extended missions, such as crewed Mars missions or orbital platforms beyond Earth, will likely require similar altitude adjustments to maintain stable orbits. Dragon's reboost test provides an opportunity to gather valuable insights into the technical requirements, fuel consumption, and thrust efficiency involved in this maneuver, which could directly influence future spacecraft design and planning.

This experiment is not only a technical demonstration but also a strategic step in expanding the ISS's operational flexibility, safeguarding its orbit, and ensuring its longevity as a hub for scientific research. Through Dragon's reboost capability, SpaceX and NASA are exploring a more sustainable approach to managing the ISS, contributing to the broader vision of reliable, long-term space infrastructure.

The reboost capability demonstrated by Dragon on the CRS-31 mission offers promising applications for future space missions, where maintaining orbital stability and adjusting altitude will be essential. This function not only supports the International Space Station (ISS) but also sets the stage for more autonomous space station maintenance and new possibilities for future orbital platforms and deep-space missions. By developing reboost capabilities within Dragon, SpaceX and NASA are pioneering methods that could shape the way space habitats, satellites, and other long-term space infrastructure are managed.

One of the most immediate applications of reboost capabilities is within the realm of future low Earth orbit (LEO) space stations. As the ISS ages, plans are being explored for commercial space stations that may eventually succeed it. These next-generation stations will likely benefit from autonomous or remotely-controlled reboost systems, allowing them to maintain a stable orbit without frequent reliance on external spacecraft. By equipping commercial platforms with the ability to perform regular reboosts, space agencies and private operators could minimize the operational risks associated with orbital decay, ensuring that these new stations remain functional and accessible for longer periods.

Reboost capabilities also hold potential for spacecraft beyond Earth orbit, where even slight gravitational forces or solar radiation pressure can alter a spacecraft's trajectory over time. For instance, on a long-duration mission to Mars or a deep-space exploration mission, periodic trajectory adjustments may be necessary to stay on course. Having a spacecraft equipped with reliable, fuel-efficient reboost systems could prevent cumulative drift from impacting mission objectives. The data gathered from Dragon's reboost tests on the ISS will be invaluable in designing propulsion systems for these missions, offering insights into the specific fuel requirements, engine configurations, and navigation systems that optimize orbit maintenance in various space environments.

In the context of the ISS, reboost technology will also play a vital role in NASA's future deorbit plans. The ISS is expected to remain operational until around 2030, after which NASA plans to retire it and safely guide it into a controlled deorbit. To achieve this, the station will need to be gradually lowered in altitude until it reaches a point where Earth's atmosphere can naturally pull it into re-entry. Developing robust reboost and deorbit capabilities within existing vehicles, like Dragon, gives NASA a new tool to adjust the ISS's altitude strategically as it prepares for deorbiting.

For the deorbit process, NASA has also commissioned a specialized U.S. deorbit vehicle, which will be equipped with enhanced propulsion systems specifically designed for this task. However, having reboost-capable spacecraft available, such as Dragon, can provide essential support in lowering the ISS in controlled stages, helping to ensure that the station's descent remains on target. These controlled altitude adjustments will mitigate the risk of uncontrolled re-entry and ensure that any remaining debris falls over designated areas in the South Pacific Ocean.

Dragon's reboost experiment, therefore, has far-reaching implications that go beyond its immediate role in maintaining the ISS. It is a step forward in advancing sustainable orbital maintenance, with applications that span from new space stations and deep-space exploration to the responsible retirement of aging satellites and habitats. By building expertise in reboost maneuvers, NASA and SpaceX are laying the groundwork for a more resilient and adaptable approach to space operations, essential for humanity's future expansion into space.

Chapter 8: The Role of Teamwork and Mission Control

The partnership between SpaceX and NASA is a model of collaboration in modern space exploration, blending the agility and innovation of a private company with the deep expertise and resources of a national space agency. This alliance has been instrumental in advancing the commercial space industry and transforming space access through projects like the Commercial Resupply Services (CRS) program. SpaceX's role in CRS, along with its Commercial Crew Program missions, has enabled NASA to focus on ambitious goals beyond low Earth orbit while benefiting from reliable, cost-effective transportation to the International Space Station (ISS). The CRS-31 mission exemplifies this collaboration, as NASA and SpaceX seamlessly coordinate logistics, payload integration, launch operations, and ongoing mission support to ensure the mission's success.

The two organizations work together at every stage, from mission planning to post-mission analysis. NASA's involvement typically begins well before the launch, where they define mission objectives, select scientific payloads, and coordinate with SpaceX on the spacecraft's configuration to meet those needs. SpaceX brings its engineering expertise to bear, designing the Falcon 9 rocket and Dragon spacecraft to fulfill NASA's specifications, including innovations like Dragon's reboost capability and its autonomous docking systems. This collaboration extends into the final stages of preparation, as both teams work closely to synchronize their operations, sharing data, conducting simulations, and ensuring all systems are "go" for launch.

Once the mission is underway, mission control teams from both organizations play a crucial role in monitoring and supporting every step of the journey. SpaceX's mission control center in Hawthorne, California, and NASA's mission control center at the Johnson Space Center in Houston, Texas, remain in constant communication, coordinating each detail to ensure mission objectives are met. Communication protocols are established so that each team knows who is responsible for specific tasks, and both centers share real-time telemetry data from the Falcon 9 rocket and Dragon spacecraft. This constant data exchange enables the teams to track altitude, speed, fuel levels, and overall system health, ensuring the mission stays on its planned trajectory.

Key personnel within each mission control center manage this collaboration, including flight directors, propulsion engineers, guidance and navigation specialists, and payload coordinators. At NASA, the flight director leads the team in Houston, overseeing all aspects of the ISS and its visiting spacecraft, while SpaceX's mission director in Hawthorne manages the Falcon 9 and Dragon. These individuals and their teams communicate seamlessly through secure communication lines, providing updates, responding to status changes, and sharing critical decisions.

Real-time problem-solving is another essential aspect of mission control's role. With missions as complex as CRS-31, unexpected issues may arise, whether they involve small trajectory adjustments, equipment malfunctions, or weather conditions. Both teams rely on their experts to analyze any anomaly immediately, and if necessary, implement a coordinated response to ensure the mission can continue safely. The partnership's strength lies in the ability to share expertise, which speeds up response times and increases the chances of resolving issues effectively.

Mission control's continuous oversight extends from launch to docking, through all phases of Dragon's journey to and from the ISS. NASA and SpaceX work as a single unit, aligning their procedures and resources to anticipate each step, adapt to any changes, and confirm that every maneuver is executed with precision. This collaborative approach underscores the trust and shared responsibility between SpaceX and NASA, each organization bringing unique capabilities and perspectives that contribute to mission success.

Through their combined expertise and clear, structured coordination, SpaceX and NASA achieve a level of reliability and efficiency that sets new standards in space exploration. This partnership not only enhances ISS resupply missions but also serves as a foundation for future endeavors, including the Artemis lunar program, Mars exploration, and beyond. Together, they are forging a pathway toward sustained human presence in space, demonstrating that cooperation between public and private sectors is key to pushing the boundaries of what humanity can achieve in the cosmos.

The CRS-31 mission, like any complex space endeavor, faced a series of unique challenges that required adaptability, teamwork, and innovative problem-solving from both SpaceX and NASA teams. These challenges spanned technical, logistical, and environmental hurdles, each of which demanded precise coordination and quick thinking to ensure mission success. Through a collaborative approach, SpaceX and NASA effectively addressed each obstacle, demonstrating the resilience and ingenuity needed to operate in the unpredictable environment of space.

One of the primary challenges in any ISS resupply mission is ensuring the safe and secure transportation of sensitive scientific experiments and crew supplies. For CRS-31, some of the scientific payloads were highly sensitive to environmental factors, such as temperature and radiation exposure, requiring specific storage conditions throughout their journey. SpaceX and NASA's engineering teams worked together to create customized storage configurations within Dragon's cargo sections, implementing climate control measures to maintain stable conditions from launch to docking. Additionally, each experiment was carefully packed and secured, ensuring stability against the vibrations and forces experienced during launch. By fine-tuning these storage and protection solutions, the team safeguarded the integrity of the scientific payloads, ensuring that they arrived at the ISS ready for immediate use.

Another challenge emerged in the timing and trajectory calculations needed for the reboost experiment, which involved Dragon adjusting the ISS's orbit. Performing a reboost with Dragon's thrusters was a new test, one that required precise modeling and calculations to prevent any unintended impact on the station's delicate systems. The teams had to work through multiple simulations to confirm that the reboost maneuver could be executed smoothly without straining Dragon's fuel reserves or impacting the stability of its orbit. NASA and SpaceX engineers collaborated closely, running real-time diagnostics and testing all propulsion systems before executing the reboost maneuver. Thanks to this careful planning and redundancy in monitoring, the reboost was executed successfully, marking an important milestone in Dragon's capabilities.

Weather conditions, always a variable in spaceflight, posed another challenge, particularly during the final days of the countdown. For the launch to proceed, meteorologists from both SpaceX and NASA tracked cloud cover, wind speeds, and lightning risks, any of which could have delayed liftoff. With close communication between launch site controllers, mission planners, and meteorologists, the team conducted continuous weather assessments to determine whether the conditions would allow for a safe launch. Fortunately, the teams were able to find a launch window that met safety requirements, allowing CRS-31 to proceed on schedule. This proactive approach to weather monitoring allowed for quick decision-making that kept the mission on track.

In addition to environmental and technical challenges, the coordination of multiple mission control teams added another layer of complexity. With mission control centers in Hawthorne, California, and Houston, Texas, overseeing different aspects of the mission, clear communication was paramount. SpaceX and NASA utilized advanced communication protocols and shared real-time telemetry data, ensuring both teams could respond swiftly to any issue and remain synchronized. This structured coordination facilitated seamless responses to in-mission adjustments and ensured that both organizations worked as a cohesive unit, minimizing risks and maximizing efficiency.

Finally, as with every ISS mission, ensuring a safe docking sequence posed both technical and procedural challenges. Any misalignment or irregularity in Dragon's approach could have risked collision or jeopardized the integrity of the docking port. Dragon's navigation systems performed autonomous adjustments to fine-tune its orientation, but SpaceX and NASA teams remained on standby to intervene if necessary. By closely monitoring Dragon's approach and maintaining precise communication protocols, the teams ensured a smooth and successful docking, ultimately securing Dragon to the ISS without issue.

The CRS-31 mission is a testament to the collaborative power of SpaceX and NASA in overcoming challenges through teamwork, creativity, and detailed planning. Each hurdle was addressed with a mix of technical expertise, real-time problem-solving, and a proactive approach to risk management. This adaptability not only ensured the mission's success but also strengthened the foundation for future collaborations, paving the way for even more complex missions in the years to come. Through careful planning and the shared goal of mission success, SpaceX and NASA demonstrated the resilience and ingenuity required to overcome the unique challenges of space exploration.

Chapter 9: A Successful Return to Earth

As the CRS-31 mission neared completion, SpaceX and NASA teams began the careful process of preparing Dragon for its return to Earth. This stage required meticulous planning, both for the spacecraft's re-entry sequence and for the organization of the cargo that would be transported back. The crew aboard the International Space Station (ISS) played a key role in this process, systematically loading Dragon with scientific samples, experimental results, and equipment that needed to be returned for analysis, repairs, or refurbishment. Given the delicate nature of many items, each piece of cargo was carefully packed, secured, and organized to ensure stability and protection during re-entry.

One of the first steps in the return process involved categorizing and safely stowing scientific samples. Experiments conducted in microgravity, such as biological samples, material tests, and environmental studies, needed special handling to preserve their condition. The ISS crew packed these items with specialized storage containers, including refrigerated or temperature-controlled units for samples that required stable conditions. Additionally, any equipment used on the ISS that required evaluation or maintenance on Earth was secured within Dragon's cargo sections. By strategically organizing the cargo, the team maximized space and ensured efficient unloading upon splashdown.

Parallel to the cargo preparations, SpaceX and NASA developed the re-entry plan, which accounted for the ideal descent trajectory, re-entry timing, and landing zone in the ocean. Dragon's return sequence is designed to protect its contents and minimize the stress on the spacecraft. The re-entry plan was carefully timed to ensure favorable atmospheric conditions and a precise splashdown in a pre-designated recovery area. SpaceX mission control worked closely with NASA to establish the exact timing for undocking, re-entry, and descent, coordinating with recovery teams positioned near the splashdown zone to await Dragon's arrival.

When Dragon was fully loaded and all systems checked, it detached from the ISS with a final farewell from the crew. The spacecraft then initiated a series of controlled burns to slowly distance itself from the station and position itself for re-entry. The descent sequence began with Dragon reorienting itself and performing a final burn to enter the Earth's atmosphere. As it descended, Dragon experienced the intense heat of re-entry, protected by its heat shield, which deflected the extreme temperatures caused by atmospheric friction.

Once it passed through the most intense phase of re-entry, Dragon deployed a series of parachutes in stages to slow its descent further. First, small drogue parachutes stabilized the vehicle, followed by larger main parachutes, which brought Dragon down to a safe speed for splashdown. The parachutes ensured a gentle descent, allowing the spacecraft to land softly in the ocean, where recovery crews were ready to secure it and bring it aboard a recovery vessel.

Upon splashdown, SpaceX's recovery teams quickly moved to retrieve Dragon, securing it and transporting it to the recovery ship. Once on board, the priority was to retrieve and stabilize the scientific samples and sensitive equipment, as certain materials needed rapid transport to specialized facilities for analysis. These items were transferred to cold storage or other secure containers to maintain their integrity, ensuring they reached researchers in the same condition as when they left the ISS.

Dragon's return process highlighted the meticulous coordination between SpaceX, NASA, and the recovery teams, each step carefully designed to protect and preserve the scientific gains of the mission. The successful splashdown and retrieval of Dragon marked the official completion of CRS-31, as the mission's cargo reached Earth safely, ready to contribute to scientific discovery and support future space operations. Through a combination of advanced technology, careful planning, and dedicated teamwork, Dragon completed its journey full circle, reinforcing its role as a reliable cargo carrier for both the journey to and from the International Space Station.

Once Dragon safely returned to Earth and its cargo was transported to the appropriate facilities, SpaceX and NASA teams began the detailed process of post-mission analysis. This phase is crucial for understanding the mission's successes, identifying any issues, and gathering insights that will inform future resupply missions. The data analysis involved a comprehensive review of both Dragon's performance and the outcomes of the experiments conducted on the ISS, as well as an evaluation of the innovative reboost test, which was one of the mission's key experimental objectives.

One of the first priorities in post-mission analysis was the examination of data from Dragon's reboost maneuver. As a new capability tested during CRS-31, the reboost maneuver data provided valuable information about the efficiency and effectiveness of using Dragon's thrusters to adjust the ISS's orbit. SpaceX and NASA engineers closely reviewed telemetry from the reboost event, including fuel consumption, thrust output, and trajectory adjustments. This information helped verify the calculations made prior to the maneuver and offered a real-world baseline for refining future reboost capabilities. Engineers could assess any discrepancies between the predicted and actual performance, enabling them to make improvements in thrust efficiency, fuel management, and control systems for future missions.

The reboost data also contributed to a broader understanding of how SpaceX vehicles might play a role in station maintenance going forward. For instance, if Dragon can consistently perform altitude adjustments, it provides NASA with more flexibility and resilience in maintaining the ISS's orbit, reducing reliance on other spacecraft traditionally used for reboosts, such as Russia's Progress vehicle. This data not only supports the longevity of the ISS but also informs planning for future orbital platforms, as reboosting will be critical for the stability of any long-term infrastructure in space.

In addition to reboost analysis, SpaceX and NASA reviewed all telemetry data related to Dragon's journey, from launch and docking to re-entry and splashdown. This data is essential for identifying areas where performance exceeded expectations and where improvements could be made. Engineers reviewed Dragon's thermal protection system, parachute deployment, and structural integrity under re-entry conditions. Analyzing these aspects allowed both teams to refine Dragon's design and operational protocols, ensuring that future missions maintain or exceed current standards for safety and efficiency.

The analysis extended to the scientific experiments and cargo delivered by Dragon. Researchers quickly accessed and analyzed experimental data and samples that had been stored aboard Dragon, especially those affected by the unique conditions of microgravity. These experiments included studies on solar wind, arctic moss growth, and materials exposure, each providing insights that could have applications in both space and Earth-based sciences. The timely analysis of these samples enabled scientists to draw conclusions about the effects of space conditions on biological and physical systems, with findings that may guide future space-based research and inspire technologies on Earth.

This post-mission phase also involved strategic planning for upcoming resupply missions. Based on insights gained from CRS-31, NASA and SpaceX could optimize cargo handling, storage solutions, and mission timelines for future flights. The teams evaluated cargo packing configurations, storage container efficacy, and the overall loading and unloading process on the ISS to determine if any adjustments could streamline operations. Each resupply mission serves as a learning opportunity, allowing SpaceX and NASA to enhance mission logistics, improve crew support, and increase the reliability of scientific payloads.

Through detailed post-mission analysis, SpaceX and NASA continue to refine their collaborative approach, ensuring that each successive mission builds on the success of its predecessors. The CRS-31 mission's data and outcomes lay a foundation for enhancing future resupply missions, advancing ISS maintenance techniques, and expanding scientific research in space. With each analysis, both organizations take another step toward more efficient, innovative, and sustainable space operations, underscoring their shared commitment to pushing the boundaries of human knowledge and exploration.

Chapter 10: Reflections on Space Exploration and Future Implications

Resupply missions have come a long way since the early days of space exploration, evolving from basic delivery operations to sophisticated, multi-purpose journeys that support a wide range of scientific, logistical, and technological objectives. In the early years of space exploration, resupply missions were relatively infrequent, highly specialized, and conducted exclusively by government-operated agencies, such as NASA and Russia's Roscosmos. These missions carried only the essentials needed to sustain crews for short-duration missions, as the technology and infrastructure to support long-term habitation in space had yet to be developed. The complexity of launching cargo into space with the precision required to dock with orbiting stations was an incredible feat, but with limited technology, these missions were constrained in their capabilities.

With the establishment of the International Space Station (ISS) in the late 1990s, the need for regular, reliable resupply missions became essential to support long-term human presence in space. The ISS required not only basic life support supplies but also tools, spare parts, and scientific instruments for the experiments conducted onboard. Over time, the focus of resupply missions expanded beyond sustaining life to facilitating advanced research in biology, physics, and materials science that could only be performed in microgravity. This shift from purely logistical missions to resupply operations that actively support research marked a significant milestone in the evolution of space resupply.

SpaceX's entry into the space industry introduced a new era for resupply missions, bringing innovation, cost-effectiveness, and reusability to space logistics. Through its Commercial Resupply Services (CRS) missions, SpaceX transformed the economics and flexibility of delivering cargo to the ISS. The Falcon 9 rocket and Dragon spacecraft, both developed and operated by SpaceX, have become mainstays in NASA's resupply program. Their ability to carry both pressurized and unpressurized cargo, autonomously dock with the ISS, and safely return materials to Earth makes them unique assets in space exploration. Perhaps most importantly, SpaceX's reusability model, exemplified by the reusable Falcon 9 first stage, has dramatically reduced the cost of space missions, making regular access to space more sustainable.

Beyond Earth orbit, SpaceX envisions a future where humanity is a multi-planetary species, a vision that centers on Mars colonization and deep-space missions. SpaceX's long-term plans include developing the Starship spacecraft, a fully reusable launch and transport system capable of carrying large crews and heavy cargo far beyond the reach of current spacecraft. Starship is designed with interplanetary travel in mind, aiming to support crewed missions to Mars, lunar bases, and even deep-space exploratory missions. Equipped with life-support systems, radiation shielding, and spacious interiors, Starship represents a significant leap forward in designing spacecraft for extended human habitation in the harsh environment of space.

The company's vision for Mars colonization involves establishing a self-sustaining city on the

Red Planet. SpaceX aims to develop systems that can transport people, supplies, and even essential infrastructure—such as habitats, power generation systems, and resource extraction equipment—to support life on Mars. Through a series of preparatory missions, SpaceX plans to send cargo missions to Mars first, delivering essential resources and setting the groundwork for human arrival. Following these initial missions, the ultimate goal is to establish a thriving human settlement that can grow and sustain itself over time, a vision that hinges on the reliability and capacity of reusable spacecraft.

In addition to Mars colonization, SpaceX's vision encompasses broader goals for deep-space missions. The development of refueling stations in orbit, advanced propulsion technologies, and autonomous systems are all part of SpaceX's strategy to make space travel as routine and sustainable as possible. By creating spacecraft that can be refueled and reused in space, SpaceX aims to enable missions to the asteroid belt, outer planets, and beyond, effectively expanding the horizons of human exploration. Each CRS mission that SpaceX undertakes contributes to this larger goal, as every innovation, efficiency, and lesson learned advances the knowledge and technology required for long-duration space travel.

Through its continued collaboration with NASA and its ambitious vision for the future, SpaceX is not only enhancing the efficiency and sustainability of current resupply missions but also laying the groundwork for humanity's next steps into the cosmos. The advancements brought by SpaceX signify a transformative moment in space exploration, bridging the gap between today's ISS resupply operations and tomorrow's interplanetary missions. By focusing on reusability, scalability, and ambitious goals, SpaceX is propelling space travel toward a future where regular missions to Mars and deep space become achievable milestones in the journey to make life multi-planetary.

As the International Space Station (ISS) approaches the end of its planned operational life, NASA is preparing for a future that will build on the ISS's legacy, advance low Earth orbit (LEO) research, and set the stage for long-term human presence in space. The ISS has been a hub of international cooperation and scientific innovation since it launched in 1998, allowing scientists to conduct groundbreaking research in microgravity. However, as the ISS ages, NASA and its partners are planning for an eventual transition that includes both the station's controlled deorbit and the development of next-generation space stations.

The ISS is currently expected to remain operational until approximately 2030, depending on its structural integrity, maintenance needs, and the continued safety of its systems. As it nears the end of its service, NASA has outlined plans for a controlled deorbit to safely bring the station down over a remote part of the South Pacific Ocean, known as the Point Nemo "spacecraft cemetery." To accomplish this, NASA is developing a dedicated U.S. deorbit vehicle, equipped with propulsion and guidance systems capable of gradually lowering the ISS's orbit in a controlled manner. This plan minimizes the risk of debris entering populated areas, ensuring that the ISS's descent is as safe and predictable as possible.

Alongside these deorbit plans, NASA is working closely with private industry to develop new, commercially operated space stations that will eventually replace the ISS. NASA's Commercial LEO Destinations (CLD) program, initiated in 2021, aims to encourage the creation of private space stations by providing funding and technical support to companies that demonstrate the capability to build and operate them. SpaceX, Blue Origin, Northrop Grumman, and other industry leaders are working on concepts for modular, scalable space habitats designed to support a range of activities, from scientific research and industrial production to tourism and education.

These private stations are envisioned as multi-purpose platforms, capable of carrying on the ISS's research legacy while opening new avenues for commercial enterprise in space. By outsourcing LEO operations to private entities, NASA can focus more resources on deep-space exploration missions, such as the Artemis program, which aims to establish a sustainable human presence on the Moon as a stepping stone to Mars. These commercial partnerships represent a significant shift in NASA's approach, enabling it to leverage private-sector innovation and investment to maintain a continuous human presence in LEO.

NASA's future space station collaborations extend beyond the private sector. The agency is also exploring international partnerships for new space habitats, building on the collaborative framework established by the ISS. Just as NASA, Roscosmos, ESA, JAXA, and CSA joined forces to construct and operate the ISS, similar partnerships are likely for future space stations. For instance, NASA's Gateway project, an orbital outpost around the Moon, involves collaboration with the European Space Agency and other international partners. Gateway will provide a base for lunar missions and serve as a model for international cooperation on other space platforms.

These partnerships will likely carry over into future LEO stations as well, with agencies from around the world joining forces to share knowledge, technology, and resources. Such collaborations will ensure that the lessons learned from the ISS continue to benefit global space programs, advancing scientific research and fostering cooperation.

In the near term, NASA remains committed to maximizing the ISS's scientific output. The agency is actively working to support and expand research in fields like biology, physics, materials science, and medicine, leveraging the ISS's unique microgravity environment while it is still operational. This focus on maximizing scientific returns ensures that the ISS remains a productive and invaluable asset right up to its deorbit, contributing to humanity's understanding of space and its potential applications on Earth.

The road ahead for NASA and the ISS is one of both transition and expansion. By carefully planning the ISS's retirement, promoting the development of commercial space stations, and engaging in new international partnerships, NASA is setting the stage for a vibrant future in LEO. This approach preserves the benefits of human presence in LEO, fosters a thriving space economy, and allows NASA to dedicate more resources to the next era of space exploration. As the ISS reaches its final years, its legacy will live on in the new wave of stations, platforms, and international projects that it has inspired, ensuring a continuous human foothold in space for generations to come.

Conclusion

The CRS-31 mission stands as a significant milestone in NASA and SpaceX's ongoing commitment to advancing scientific research, strengthening space logistics, and enhancing spacecraft capabilities. Through the successful delivery of over three tons of cargo, including essential supplies, scientific instruments, and experimental materials, CRS-31 contributed directly to the work of the International Space Station (ISS) and its crew, supporting both immediate needs and long-term research goals. The mission also marked a technical achievement with Dragon's reboost experiment, a test that demonstrated the spacecraft's ability to independently adjust the ISS's orbit—a new capability that offers NASA greater flexibility and resilience in station maintenance. From the seamless launch and docking to Dragon's successful splashdown and retrieval, CRS-31 exemplifies the high standards of safety, precision, and innovation that define modern space exploration.

The legacy of CRS-31 extends beyond the mission's immediate accomplishments. Each scientific experiment conducted in the microgravity environment of the ISS—from the solar wind study to arctic moss experiments and materials exposure tests—adds a new layer to humanity's understanding of space and its impact on biological and physical systems. These findings will help scientists improve technologies on Earth and develop solutions that could support life on other planets. Moreover, by successfully testing Dragon's reboost capability, CRS-31 has paved the way for future missions that could use this technology to support station operations and even deorbit the ISS when it reaches the end of its operational life. The mission underscores the value of reusable, versatile spacecraft in achieving ambitious exploration goals while reducing costs and expanding possibilities for long-term human presence in space.

Looking to the future, CRS-31 represents just one step in a rapidly evolving era of space exploration, one in which public and private sectors work hand-in-hand to push the boundaries of what is possible. In the coming decade, SpaceX's vision for Mars colonization, coupled with NASA's Artemis program for lunar exploration, promises to take humanity deeper into the cosmos. The ongoing development of SpaceX's Starship for interplanetary travel, the next-generation Lunar Gateway station, and private space stations in low Earth orbit all signal a future where space exploration is not only sustainable but also expands our presence beyond Earth in meaningful ways. These initiatives will continue to build on the foundation laid by missions like CRS-31, advancing our knowledge, testing the limits of technology, and bringing us closer to a multi-planetary future. As we stand on the threshold of these new frontiers, the successes of CRS-31 remind us that each mission is a building block toward a vision of space exploration that benefits all of humanity.

www.ingramcontent.com/pod-product-compliance
Lightning Source LLC
Chambersburg PA
CBHW070424240526
45472CB00020B/1182